NEGOTIATING YOUR CAREER IN
STRUCTURAL ENGINEERING

Er. M.A. BEG
BE, LLM, FCIArb.

Negotiating Your Career in
Civil Engineering Series.

First published in 2020 by

This edition published in 2020

© 2010 Mohammad Akhtar Beg

All rights reserved by the author of this book. No part of this book may be reprinted or reproduced, or utilized in any form or by any electronic, mechanical or other means, now known or hereafter invented, including photocopying and recording, or in any information storage and retrieval system, without permission in writing from the author.

SOME REVIEWS

The book provides a smooth walkthrough for a career development of structural engineers. It gives insight on the basic requirement and knowledge required for navigating a successful career path as a structural engineer. The book is well structured and covers the details of various training needs, information about various design consultants & companies, structural engineering software, etc. The book also has a section for entrepreneurship, which is very enlightening, and also has a chapter on preparing and presenting a CV. I strongly recommend this book for all civil engineers, students of Civil Engineering and specifically those who want to pursue a career as a structural engineer.

Dr. Anubhav, (Ph.D. IIT Kanpur)
Add. General Manager
NTPC Ltd., India

Choosing and pursuing a career can be quite difficult, as the decision taken from the options known & available in today's times will decide the rest of the life ahead. Therefore, one has not only to choose a field but the future.

Having been a civil engineer for the last more than 50 years working with lot of structural engineers on different projects across the country & abroad, I have seen the changes in the field of structural engineering & adventures in this field. In line to define the role of a structural engineer and to turn someone's dream in to reality with a lot of space & opportunities for creativity, I am delighted to see a guide in the form of this book where in the author has done a 'manthan' to provide to aspirants in the field of Structural Engineering as a career.

B. B. Kumar
Executive Director in NBCC
Executive Vice President in NCC Ltd
President Construction in Dhoot Group

As a reader I went through this book and noted that many features are included, like starting from setting up your goals to fulfilling the goals. The book sheds light on how to go about building a career in this field which I think will definitely be useful to young engineers who really want to move forward in their careers. I can conclude my review after reading this book by saying that, all young structural engineers can go through this book to become an efficient structural engineer in society.

Prof Shiv Shankar KM
Dept. of Civil Engineering, ACS College of Engineering

The book is a must buy for structural engineers as it guides them how to pursue a career path. I found this book equally useful for freshers as well as for midlevel structural engineers.

Vimal Kak
General Manager
My Home Enterprises, Hyderabad

As an academician I went through the book and found it to be very *practical. It leads the young structural engineer through a path which they can follow to start and move up in their career. A perfect book for young engineers who are aspiring to become structural engineers.*

Sachin Amarnath
Academician

I found this book full of practical information for structural engineers and it contains solutions to real time issues being faced by young structural engineers in their career. A good read for young engineers.

Prof. Arjun K
Department of Civil Engineering
BMS College of Engineering

ACKNOWLEDGEMENTS

I would like to sincerely thank the following people for their continuous support and guidance. This book has been a result of their continuous cooperation and encouragement.

Dr. Anubhav (Ph.D. IIT Kanpur), Add. General Manager, NTPC Ltd., India.

Mr. B.B. Kumar, Executive Director in NBCC, Executive Vice President, NCC Ltd., President Construction in Dhoot Group.

Mr. Haries Salim, Project Manager, Towell Construction Co LLC, Muscat.

Mr. Vimal Kak, General Manager, My Home Constructions.

Prof Shiv Shankar KM, Asst Professor, ACS College of Engineering.

Prof Arjun K, Department of Civil Engineering, BMS College of Engineering.

Mr. Sachin Amarnath, Academician.

Mrs. Rohini S, Asst Professor, K.L.N College of Engineering, Chennai.

TABLE OF CONTENTS

Preface	xi
Introduction to the Series	xii
Introduction to This Book	xv
The Aim of This Series	xvii

CHAPTER-1: CONTEMPLATING WHETHER TO PURSUE A CAREER IN STRUCTURAL ENGINEERING — 1

1.1 Introduction	1
1.1.1 Reasons for Becoming a Structural Engineer	2
1.1.2 Where can You Start Your Structural Engineering Profession?	3
1.1.3 Conclusion	5

CHAPTER-2: PREPARING FOR WORK — 6

2.1 Introduction	6
2.1.1 Self-Awareness	6
2.1.2 Set your Personal Goals	8
2.1.3 The Strength and Importance of Communication	9
2.1.4 The Importance of Knowing English	10
2.1.5 Working in Corporates vs Small Firms	11
2.1.6 Corporates	11
2.1.7 Small Firms	12
2.1.8 Conclusion	13

CHAPTER-3: THE GUIDED PATHWAY	**15**
3.1 Introduction	15
3.1.1 The Route to Success	16
3.1.2 Conclusion	25
CHAPTER-4: SOFTWARE USED IN STRUCTURAL DESIGN	**26**
4.1 Introduction	26
4.1.1 List of Software Commonly Being Used for Structural Design and Analysis	28
4.1.2 Conclusion	33
CHAPTER-5: PROFESSIONAL INSTITUTIONS AND ASSOCIATIONS	**34**
5.1.0 Introduction	34
5.1.1 What Are Professional Institutions or Organizations?	35
5.1.2 Benefits of Joining a Professional Institution or Body. There are a lot of benefits in joining a professional body as	35
5.1.3 Grades of Membership	37
5.1.4 How to Make Most of Your Membership	39
5.1.5 List of Professional Institutions	39
5.1.6 Conclusion	46

CHAPTER-6: THE IMPORTANCE OF TRAINING AND DEVELOPMENT AND LIST OF REPUTED INSTITUTES WHICH PROVIDE THESE   48

6.1 Introduction   48
6.1.2 Online Training, Development and Certificates   51
6.1.3 Other Useful Resources for a Structural Engineer   54
6.1.4 Conclusion   57

CHAPTER-7: SOME REPUTED COMPANIES AND THE JOB FUNCTIONS OF A STRUCTURAL ENGINEER AT VARIOUS LEVELS   58

7.1 Introduction   58
7.1.2 Where to Start Your Career   58
7.1.3 Some of the Renowned Companies and Firms who Offer good jobs and growth   62
7.1.5 Layout of Job Responsibilities in a Typical Design Office   71

CHAPTER-8: STARTING YOUR FIRM   75
8.1 Introduction   75
8.1.2 Why to Start Your Own Business   76
8.1.3 Carve out a Niche for Yourself   76
8.1.4 Professional Membership   77
8.1.5 Experience   77
8.1.6 Check the Statutory Requirements   78
8.1.7 Funding Your Business   79

8.1.8 Insurance Cover	80
8.1.9 Try to Have Some Clients in Hand before you Cut Loose	81
8.2.0 Some Essentials Which You May Need to Know	81
8.2.1 Start looking out for Big Contracts and Tenders	83
8.2.2 Conclusion	84
CHAPTER-9: SOME TIPS ON WRITING YOUR CV	**85**
9.1 Introduction	85
9.1.1 Some General Tips at First	86
9.1.2 The Covering Letter	89
Conclusion	91

PREFACE

"The idea of this series slowly developed as I journeyed through graduating from college to starting my own firm. I moved through various attempts and failures, not knowing what to do and how to proceed. The movement in my career was a sort of automatic; I closed down my firm; then I took a shot at trying my forte in education; and finally joined the construction field, working hard and achieving the level of a Project Manager. As I mentioned, the rise was due to my experience and the hard work I put in. As I continued in my pursuit of the ultimate goal, I gradually realized the meaning and importance of career planning, and recognized that there was an information link missing in most of our young graduates' education which leaves them confused once they successfully pass out from their colleges. I have tried to somehow fulfill this missing link by writing down a series of books which aim at guiding the civil engineering graduate to creating a path for building a successful career."

The Author, Mohammad Akhtar Beg, is a civil engineer with 26 years of rich field experience in project management and project execution. His forte has been structural designing, planning, quantity survey and contracts. The author has designed several structures and has always been connected to this field and has also been instrumental in designing and handling structural problems in many projects.

INTRODUCTION TO THE SERIES

Why I Wrote This Book.

"Life is full of uncertainty. Risk is a part of every choice. The threat of failure looms around every corner. If you're not careful, these hazards can stop you. You can get stuck before you start, not knowing what to do."

Jeff Goins

I decided to write this first book out of a series of books on careers in civil engineering, after realizing a hard fact through my own journey in this field. The fact that a young engineer does not understand the significance of planning a career until around 10 years post graduation. This is true for nearly all of us engineers except maybe a selected few. When a youngster steps out of their college as a qualified engineer, full of accomplishment and dreams and hopes for the future, they are too occupied with the success and are also too happy and relieved to have attained a commendable triumph. Apparently, thinking about a career immediately after accomplishing this big feat is not the first priority at this young age. It is sometimes the sense of freedom and ecstasy which takes them to a new high and clouds their minds for some time.

I have decided to write a series of books on civil engineering careers out of which this is the first book I have written. Further books on the series will cover other disciplines of civil engineering.

It is in the aftermath of this ecstasy that young graduates find that he has little knowledge of how to take the first step in his career. This guide aims to remove the deliberation of the unsure and unknown future and the lack of awareness of what to do next and of what needs to be done to kick start a career.

Not knowing what to do is a dangerous situation in itself. I have seen young engineers struggle to decide what to do. Your teachers in your colleges are good academicians but unfortunately seldom know about the practicalities and challenges which you will face and should brace for, in your career in the industry. They are a different class, more focused on the research and academic aspect of engineering and have great capabilities and expertise in those areas. Unfortunately, this kind of expertise may not help you in getting a job in the industry.

Through this book, I have attempted to guide you on how to take some tangible, concrete, and quantifiable physical actions which can help you start your career and move up in life. I have tried to make it as simple as possible and laid down the precise steps which you may need to take from the start till you become a renowned personality in your field. This guide is not meant to be an inspirational or spiritual guide for your inner self; on the contrary, it gives advice, on the real world and talks about the measurable actions you can take to promote your career.

This book also addresses problems that young engineers like you face in their midcareer levels where they may find themselves to be stuck in a position wanting to move up in their careers, but unable to do so due to the absence of having no options. In such cases, it is common for engineers to keep changing organizations in the anticipation of a better position and job satisfaction. This book sets out some very practical advice that you can follow to become more confident and thus take some balanced and tangible actions.

The book will also guide you on how to take calculated moves and when to take them, which will ultimately result in a better future for you and will save time which you might unnecessarily spend in planning your otherwise unknown moves. It will help you gain confidence and knowledge of what you can expect and the numerous requirement of this field and guide you to build a successful profession.

I have written this book through the learnings I have gained from my experience in building my career and how I have grappled initially and in the midst of my career, sometimes struggling to find answers which were seldom available. My intention is to write as simply as possible so that the reader can comprehend their own situation and act accordingly.

INTRODUCTION TO THIS BOOK

"No business can succeed in any great degree without being properly organized."

James Cash Penny

This book is intended to guide your transition from a civil engineering graduate to a **professional Structural Engineer** by imparting the required command and confidence which is needed in the industry, so that you are braced and ready to face the cut-throat competition in the present-day job market. The book also addresses the opportunity and the necessary steps required for becoming an entrepreneur in this field.

The Statistics:

Notwithstanding the economic crisis induced by the Covid-19 pandemic, the global civil engineering market is anticipated to witness tremendous growth in the upcoming years which is attributed to the rise in disposable income coupled with technological advancements in civil engineering projects in the overall construction industry. A high demand for residential buildings and infrastructure, in emerging economies such as China, India and other South Asian countries, is further expected to generate impressive growth, in the civil engineering market in the next few years. An increase in focus on bringing more innovation with regards to construction materials is one of the rising trends in the construction industry, which is expected to fuel the demand for civil engineering designs.

It is projected, that to improve infrastructure, large investments by the government sector is anticipated to boost the demand for the global civil engineering market by the end of 2023. The impressive growth in the real estate sector in the developing countries of Asia-Pacific such as India and China, will also enhance the demand for the civil engineering market. Hence, the global civil engineering market is expected to witness a CAGR of around 5.4% over the forecast period. However, a lack of skilled labor can be a major restraint for the civil engineering market.

The US Bureau of Labor Statistics has projected a growth of 6% in civil engineering jobs from 2018 to 2028, which is about as fast as the average amongst all occupations. As infrastructure continues to age, civil engineers will be needed to manage projects to rebuild, repair and upgrade the infrastructure of a country.

Also regarding the salary of a civil engineer, there is some impressive data. Civil Engineering was one of the most in-demand jobs with an average salary of $59,000 in the early stages of the career to $96,000 in the later stages of the career. This figure may vary from country to country a bit, but in general civil engineering jobs may still rank among the top jobs in most of the countries.

THE AIM OF THIS SERIES

The Civil Engineering field in itself is a very vast field, which includes tasks ranging from design planning and construction of buildings, infrastructure and the like. It gives a lot of opportunities to a civil engineer of your caliber to learn and to work in these diverse areas. As there are a lot of divisions a civil engineer can work in, it sometimes becomes confusing for a fresh civil engineer to decide as to which field of work to specialize in. You should understand that civil engineering offers a lot of opportunities in which an engineer could gain experience and move on to become an expert in that field. This book aims to guide a fresh civil engineer on how to start and pursue a career in structural engineering. It also aims to guide civil engineers currently in their mid-career levels. I have come across civil engineers working at these levels who seem to be confused as what to do next to advance their career to the next stage. You can say that they suddenly have hit a plateau in their career graphs after advancing considerably. If you are one of them, then this book will guide you on how to explore the various means to enhance your skills and knowledge, through which you can overcome this phase and advance your career again to newer heights. With these skills acquired, you can become an internationally accepted engineer in your field; and this can also pave the way for you to start a job or enterprise in the national or international arena.

The civil engineering field is a branch of engineering which evolves regularly and hence engineers associated with this filed need to keep abreast with the development of recent technologies. It is a field so vast that it is difficult to be an expert in multiple disciplines; instead, you as a civil engineer should focus on one discipline which interests you; and should be steadfast in the pursuit of your goals. Dale Carnegie, a famous author said "People rarely succeed unless they have fun in what they are doing.", so for success, it is inevitable that you should have fun in what you do and, at the same time, you should be ready to embrace and learn and adapt with the very rapid changes that take place in this industry.

After reading this guide, a civil engineer either fresh or experienced, will be able to identify whether they are interested in becoming a structural engineer; acquire knowledge of how to start and develop their career; and keep abreast with the best renowned and acknowledged industry practices in the world; so that they have an edge above the others and get noticed in the industry.

However, as it is well said by someone that "There is no elevator to success, you need to take the stairs"! Nurturing a career may take some time and by reading and following this guide, the time spent on developing a career of an engineer will be well invested.

CHAPTER-1

Contemplating Whether to Pursue a Career in Structural Engineering

1.1 Introduction

Becoming a structural engineer can be one of the most exciting things in your life, it is something that is intellectually challenging at times, but massively rewarding too! Of course, deciding to become a structural engineer is a very important decision, as it may be a decision which you may have to live with, throughout your life. After stepping into the shoes of a structural engineer, it may be difficult for you to remove them and assume a new role. In this context, this chapter is meant to give you guidance and help you with your decision to take up a career as a structural engineer. Subsequent chapters will support this decision of yours and provide information about the process of becoming a respectable and worthy structural designer. Also, on further reading this guide, you will learn about the steps needed to be taken to attain a successful career and the various career opportunities available to you.

1.1.1 Reasons for Becoming a Structural Engineer

There are some principle reasons which I have listed below, and should any of them apply to you, I would strongly suggest that you take up a career in structural engineering. These are listed below.

1. **You are interested in further education and training:**

    If you have a liking for continuing your education to a Master's level, and then maybe to a doctorate, and you desire to learn and develop yourself and are fond of this, then you can move forward and take up this field.

2. **You are fascinated with the large structures, buildings, and bridges that you have seen**

    If you tend to be fascinated and love large buildings and structures and wonder how these are erected and stand so tall, and want to know how to build these.

Dr. Torbjörn Wigren, a Senior Engineer at Ericsson AB, once commented "My career goal was to obtain a challenging engineering position that focused on design and innovation rather than on simple implementation of the idea of others." To achieve his goal, Dr. Torbjörn Wigren went on to obtain a doctorate. Likewise, if you also strive for a position which is focused on design and innovation then you can opt for this field.

3. **You are interested in science and technology:** You will know if you belong to this category, if you try to look at structures deeply and you delve into these and have a desire to know more about them.
4. **You have the confidence of building structures:**

    If you possess the confidence that you can become a structural engineer and build structures and have the desire to be a part of the vast designer's fraternity then this field is for you.

1.1.2 Where can You Start Your Structural Engineering Profession?

Although I have addressed this in detail in Chapter 3, I bring up this topic here merely to discuss this at a broader level. Here we will discuss the initial areas where you can work as a structural engineer in India or in the international arena.

5. **Large Private Ltd Corporate firms:**

    Large corporate firms offer the best employment opportunities for a fresh structural engineer. I have detailed the names and the types of firms and you can go through them in Chapter 7.

6. **Government Undertakings:**

    Government undertakings in India and worldwide have not

been very active in recruitment in the recent past, primarily due to the focus that the respective governments have for outsourcing these jobs rather than getting them done through their own in house departments. Added to this is the fact that, there are fewer opportunities available. Moreover, governments have their own procedures for employment and these differ from nation to nation. You can explore these by understanding the procedures and tests which you may need to take up. Even though careers in a government undertaking like Mecon Limited, NTPC, ONGC, etc. can be thought of, but these are very rare and the opportunities of recruitment are far too apart.

7. **Small Firms:**

Small firms also provide good career opportunities, but they have their own downsides. I have discussed in detail about the advantages and disadvantages of working in corporate and small firms, which is detailed in Sec 2.1.7 for a quick read.

8. **Private Consultant or Entrepreneurship:**

Becoming a private consultant or opening your own consulting firm is always a good idea; but this should be done after you gain substantial experience in your work. I have detailed this work in Chapter-8.

1.1.3 Conclusion

In this chapter, I have described how you can decide and contemplate on developing yourself into a professional structural engineer. This decision is very important at an early stage in your career, since it will be difficult for you to change careers later on. In terms of employment opportunities, I have given you an overview of the areas where you can start your career. Remember structural engineers seldom tend to move around several fields, that is they can become experts in buildings and remain there throughout their lives, However, it is also possible that you may be tasked to design buildings and infrastructure work also. This will increase your experience. Nevertheless, it is better to choose a field and become an expert in that, while remaining agile and ready to take up work in other areas too.

CHAPTER-2

Preparing for Work

2.1 Introduction

You have graduated from your college and are now equipped with a degree. You are brimming with confidence and are ready to take up work. This chapter will guide you on how to prepare yourself for work, through different techniques and strategies which you can adopt while preparing yourself to take on the world.

2.1.1 Self-Awareness

Self-awareness is your ability to focus on yourself and direct your actions, thoughts, emotions, and dos and don'ts to align with your internal standards. Tasha Eurich writes in her post "Research suggests that when we see ourselves clearly, we are more confident and more creative. We make sounder decisions, build stronger relationships, and communicate more effectively. We're less likely to lie, cheat, and steal. We are better workers who get more promotions. And we're more-effective leaders with more-satisfied employees and more-profitable companies."

Self-awareness is of two types - **internal** and **external** self-awareness. The **internal** self-awareness represents how clearly we see our own values, passions, aspirations; fit into the environment, reactions, etc.; and this type of awareness is associated with a higher job satisfaction, personnel control, happiness. But this is also negatively associated with depression and stress.

External self-awareness is basically how we understand how other people view us with respect to the above factors. People who have a high self-awareness are more skilled at showing empathy and accepting perspectives from the other's point of view.

Being self-aware about oneself is important as it will give an insight into your inner strengths and weaknesses and how you can know and understand your character. This will help you in deciding the types of jobs you can be interested in, and how you can improve your behavior at the workplace.

There are several online self-awareness tests freely available; you can attempt these and take time to reflect on yourself. Numerous self-awareness courses are also available which may be worth taking to introspect upon yourself.

2.1.2 Set your Personal Goals

It is a good practice to set your goals about what you are planning to do with your precious life in the next say 5 to 10 years. It can be called a plan - a strategic plan - and setting your goals clearly and with a mission and vision gives you a focus to attain your ultimate aim early on. Your goals should be **SMART goals;** they should be **specific, measurable, attainable, relevant** and **time bound.** All professionals including students, young engineers, and practicing professionals, should set their goals and keep a track of their achievements.

Goal setting gives you the chance to experience the power of your imagination and keeps you on track. Setting goals involve thinking about your future, thinking about your tomorrow, so that you can start reflecting and prospecting on the goals that you would want to achieve in the near future.

Setting goals will also give you a long-term vision and motivation. It enables you to organize your time and resources in a planned manner, so that you can make most of your life. Achieving sharp and clearly defined goals will give a sense of accomplishment and pride, and you will see progress in areas which earlier would have seemed to be nothing but some meaningless grid.

2.1.3 The Strength and Importance of Communication

"Communication – the human connection – is the key to personal career success."

Paul J Meyer

Communication represents conveying meanings from one entity to the other through mutually understood language or signs. Effective communication in a workplace is not just about exchanging information. The emotion and the intentions behind the communication also need to be understood. Good communication is not only about conveying a message; it also involves listening to what's being said and making the speaker feel that they are being heard and understood.

Charles H Thornton, the co-founder of the international structural firm **Thornton Tomasetti,** in an interview said "Communication is probably more important than anything else. If a young engineer can't communicate at the point of earning a master's degree, it may be too late."

Good and effective communication is a skill which cannot be taken for granted, you start communicating first with your CV, then it moves on to your first interview and then to your job. Strong communication skills are a necessity and not just a talent.

Effective communication skills come naturally to some people, for others it is hard for them to articulate their thoughts and feelings in conversation, which at times lead to misunderstandings and conflict. Failing to communicate effectively may be detrimental for your career and you may end up losing your dream job. On the other hand, effective communication at the workplace is an essential part of business success, it keeps the team working together and increases employee engagement. Teams that do not communicate properly fail and make mistakes. The US firm **Gartner** shows that a whopping 70% of business mistakes are due to poor communication. This statistic proves how important communication is.

2.1.4 The Importance of Knowing English

I may sound a bit biased, but it is a fact that you should learn to communicate effectively in the English language. I cannot state how important this is, as all communication in a design firm is done in English - you design in English, you draw in English; and it is an accepted norm internationally that engineers in general communicate in English. If you are not adept with speaking and writing in English, it is now that you should start working on improving your English language skills. There are a lot of courses offered by reputed institutions like the British Council, and BBC which one can take online to increase their proficiency in the English language. In fact, if you plan to work internationally in any English speaking country, which are a lot in number, you may be asked to pass and get the required grades in IELTS or TOEFL exams so as to prove your proficiency in the English language. In fact, all engineering and construction communication at all significant levels takes place in

English around the world. In other words, English is basically the accepted language of communication for all engineers around the world.

2.1.5 Working in Corporates vs Small Firms

Once you set your foot in the industry, you will be faced with a choice of working with either large Corporate companies having a national or international presence or relatively small firms with a lesser outreach. I have outlined below the pros and cons of working with each of these types of organizations, which will help in deciding where you aim to work.

2.1.6 Corporates

Large corporations usually have a very defined and structured hiring process and you may have to go through a number of screening processes and interviews before you are handed the most desired appointment agreement to sign. Nevertheless, large corporates have very well-structured systems and processes and you will be required to take up specific roles and responsibilities only. You will be one of the many members working as part of a team at a large corporation; and if you enjoy collaborating and relying on group dynamics, then you will surely enjoy your job. Most of these corporates have training programs and development programs which prepare the employees for future roles, and you will gain benefit from these.

Facilities and perks provided by large corporates are standard and some of them are quite generous. Also, the job stability in a corporate is generally much more than in a small firm. The name of the corporate in your experience section of your CV will always give you an advantage above the others if you would like to look for a job change later on in future.

Nevertheless, working in a corporate carries with it some disadvantages too. It is generally tough to think out of the box and experiment with processes in large corporates sometimes, as these companies work on certain pre-defined policies and procedures which you may find tough to bypass. Working in a corporate means that you are not alone, you are working with a big group of people and getting ahead in this group will be tough and the competition will be hard. It may be difficult for you to outshine the others and work on other disciplines if you so desire.

2.1.7 Small Firms

You may be able to land a job easily in a small firm as their procedure for selection is much faster and less bureaucratic.

Small companies will be more ready to think out of the box, and if you are good at your work and have sufficient skills, you will be able to standout and grow fast in your role. You will have lots of opportunities to work, unlike large corporates, and your role will not be constricted within defined deliverables. In a small corporate you can juggle between different roles. If you can find some opportunity you are passionate about, you can become a driving force in the company's success and if you have the ability, you can get bigger responsibilities to handle in a short span of time.

In small firms or companies, the environment amongst the team is more or less attached...... there is a strong feeling of community and belongingness and such relationships will help you in getting mentored and provide much required guidance.

Like working in big corporates, small firms also have their disadvantages. In small firms, you may not be able to get exposure to big projects and the training and development programs will many a times need to be financed by yourself, if you desire to take them. There will be fewer chances of growing professionally, and also lesser personal benefits. Job security and stability in a small company is also low and you may be at a risk of losing your job in a short time if things start going downhill.

2.1.8 Conclusion

In conclusion, you can see that there are many pros and cons in working with either a large corporate or a small firm. However, while starting your career, I can tell you from my experience that it is worth trying to join a large corporate at the beginning, as, at the start of your career, you will be in the need of a good break and if you can put your foot through the door of a corporate at this stage, you will get the exposure you need to move upward. The more you are exposed to the environment, the more experience you get, and it is well said that what matters is twenty years of experience and not one year of experience repeated twenty times.

You can also read further in Chapter 7 where I have listed some of the most reputed companies and consultants who offer jobs to structural engineers around the world as well as in India.

CHAPTER-3

The Guided Pathway

"Theory without practice is of little value, whereas practice is the proof of theory. Theory is the knowledge, practice the ability."

Alois Podhajsky

3.1 Introduction

To pursue a career in structural engineering you will need to understand how a standard structural department or firm functions, and what are the job responsibilities of each position. This chapter sets out the information on the various roles you will assume as a structural engineer in the course of your career. Being familiar with these roles will help you plan your career and open up the course of action you may need to take in advance. This information will help you understand what to expect once you start your career.

3.1.1 The Route to Success

"The well-trodden path is not always the right path"

Tahitian Proverb.

The knowledge of what lies in wait for you is better than finding your way in the dark. This makes you more prepared to take any challenge head-on and builds up your confidence.

Here I have endeavored to share with you, how you should navigate your career as a structural engineer. You will learn about how you can start your career and continue to progress on a consistent path, growing in your career as you gain knowledge and experience.

Needless to say, to climb this ladder of success, you will be required to exhibit your knowledge, market yourself, and show to the world that you have the ability, skills, and confidence which are expected of you as a structural engineer.

The path is described as a step-by-step guide, explaining to you the significance of each step. This stepwise structure will be easy for you to understand the career path and will make you aware of the various stages that you will pass through in your career.

STEP-1..................The Start

1. Structural engineering is a field that requires knowledge of the

behaviour of structures. It is advised for those engineers who like to study further and keep progressing on a course of research in their career. If you are of a type who does not have much love for studies and research, it will be better for you to develop these if you want to be a reputed structural engineer; or you might as well opt for another field of your choice.

2. Always be open to learning and be educable; getting a bachelor's degree is not the end of your learning experience. Instead, it is the beginning of long and rewarding progress and you should always look forward to learning continuously. Taking up the books may be a bit difficult immediately after you finish your undergraduate program. Take a short break of a month or so and return to your learning table. This will take you to new heights of confidence and propel your career forward with a force which cannot be resisted by anything in the world. Remember, if you are not willing to learn, nobody can teach you; and if you are determined to learn nothing can stop you!

"Learning is not attained by chance, it must be sought for with ardour and attended to with diligence."

Abigail Adams

This will be your first step in professional development and this is where you start developing your skills and competence and begin to apply your professional judgement.

3. Opt for a Master's program in structural engineering from a reputed college. There are many colleges offering a Master's degree in Structural Engineering and Design around the world. Take these and strive to get your Master's degree. A Master's degree in this field is a necessary requirement, and you can opt to get it immediately after your Bachelor's degree or you may choose to take it after some time. Nevertheless, opting for a Master's degree is a necessity, and the sooner you get it the better.

4. In case you do not want to pursue a Master's degree immediately after you graduate from your college, you should start taking job-specific trainings in the areas of design that you want to further your career on.

STEP-2 ....Take the Step

1. Do not look for earning money at this stage. Keeping Step-1 as a prime motive, strive to join a reputed design firm. I have detailed a few of the renowned design firms and companies in Chapter-7. These companies and firms are always on the lookout for fresh talent and you can start applying to them during your Master's program. If you can demonstrate your talent and skills, you can get a position in those companies or any other firm. It is also possible that some of these companies can fund your education too! If you are unable to do so, go ahead and join a firm which works on medium projects. You have to be patient and try to learn the skills of the trade. Learn

other skills like drafting, as a structural engineer is also required to have a good understanding of drafting of details of the design and give instructions to the draftsman. Strive to learn these, as these are seldom taught in colleges.

2. You should start working on your career immediately after you complete your under-graduation course, either by joining a Master's program or by joining a firm. Do not opt for starting to design projects on your own without gaining sufficient experience and knowledge.

STEP-3............... Join the Professionals

3. If you are not a member of a professional institution, now is the time to become one! Being a member of professional institutions and societies shows your interest in the awareness of the industry and the recent trends to your prospective employer. It is a mark of professionalism which you can show off to your employers. These memberships are usually free for student and if you are a student reading this book then I would strongly suggest you to join one of the institutions I have listed in Chapter-5. These memberships offer a lot of value and will guide you on how to develop professionally and how to develop your personality as a structural designer.

4. By joining any of the listed institutions mentioned in Chapter-5, you will make valuable professional contacts and get access to a wealth of information. The other most prominent advantages

of joining a professional organization are:
a. Recognition of the member's professional status, which may include post-nominals.
b. You will stand apart from other candidates if you have a membership status of some of the reputed institutions.
c. Many institutions provide career resources that help candidates get a break in the job market.
d. You will get access to exclusive online resources that you use.
e. You can get exposure to a lot of networking opportunities by attending meetings and seminars.
f. You can traverse a regular path of professional development which will enhance your skills and make you more employable.
g. You can be up to date on information about the Industry standards and education opportunities, as these institutions often distribute newsletters that keep members informed about new statistics and best practices in the industry.
h. Mentoring programs are one of the greatest advantages that these institutions provide. You can find a mentor who can guide you through most of the challenges you face in your profession.
i. Certification Assistance is provided by many of these institutes, whereby you can get a chance to sit in a review course that helps to prepare you to pass any certification exams which may be a prerequisite in many countries for you to practice independently.

5. Joining social networks like LinkedIn will help you a lot as you can join communities of structural engineers and follow institutions that offer continuous discussions on real-time

problems in structural engineering. This will increase your knowledge and give you a feel of the market conditions and requirements.

STEP-4..................Train

1. Structural engineering is a vast field with different requirements for different types of structures. Structural engineers tend to hold on to one of the fields and continue to tread on a monotonous course; rarely trying to deviate from the trodden path. For example, a structural engineer designing buildings will develop a comfort zone around these and will try to be in a position where a similar type of work can be found. In this way you may keep on working, but by following this comfortable attitude you may end up in a plateau in your career without moving upwards. To stop this, always be on a continuous path of learning and development and never let go of any opportunity to learn. Widen your comfort zone and always be on the lookout for newer things.

2. Start taking training while working. This will increase your knowledge and count towards your continuous professional development. Courses are provided by most of the professional institutions as listed in Chapter-6. Please note that by taking these courses you may invest some money; but the money spent will be worth, as these courses will guide you to become a professional and you can gain a lot of experience in carrying out your design with approved industry methods and

can get to know what design practices are being adopted by the industry. A detail of courses and trainings is provided in Sec.6.1.2, which you should explore and learn according to your requirement. These courses will increase your knowledge and give you an international exposure of good and standard practices in design. You would be surprised at the amount of knowledge that you can gain from these.

STEP-5 Know the Good Industry Practices

3. Every engineering trade is performed with a certain acceptable set of quality standards, and this quality in your work can be achieved by learning and following internationally accepted professional design practices. All the reputed organizations follow these practices; and as the world is awesomely connected, you may want to learn to design as per accepted international standards. By learning these practices, you become accepted as a professional in your work throughout the world. Always use these good and industry-accepted practices while designing, this will show the required professionalism in your work at all stages.

You can learn these by experience and also through courses that will guide you to the best and standard practices being adopted in the industry worldwide (please refer Sec 6.1.2). These can be learnt while you are working and online which makes it much easier to pursue, and if you do so, you will have the advantage of applying these practices to your designs.

STEP-6 The Certification / Accreditation

4. Certification is a mark of professionalism in a structural engineer's career. It makes you stand out from the crowd and you start to become recognized as a professional internationally in your field. You become a more noticeable and in demand structural engineer and can be picked up by employers easily.

5. Certification breaks the roadblock which comes in one's career after about 10 years of work. During this period, you would have achieved some success and would be well settled. This is a period where you tend to slow down as you would have overcome the initial struggle in your career. This slowing down and satisfaction might at times lead you to a flat curve in your progress. By getting a certificate from a reputed institution, you can propel your career growth and start a new rise in your career graph.

6. A certification can be had while you continue to work on your job as most of these are online proctored exams. These exams will test your ability of design and other areas of structural engineering. I have provided a list of reputed institutions providing certificates for a structural engineer, and these certifications will be a further feather in your cap. I am listing some of the great benefits you will gain if you attain a certification:

- Certification serves as a means of differentiating oneself from other uncertified engineers.

- It helps in recognition of the standards that the certified engineer follows while working.

- It gives an opportunity to structural engineers to explain to others how they are distinctly qualified.

- It assures your employer and the public that you can carry out your functions safely and ethically and you possess the desired competencies in your field.

- The certification is a means to show to the world that you are an expert in your field and possess the necessary knowledge, and it is an immediate recognition of your profession and your capabilities.

STEP-7 THE LICENSE

7. Start your procedures for becoming a licensed structural engineer as some of the countries require that a structural engineer should have a license before they are allowed to design structures independently. After getting licensed, you can start designing on your own, and can also start your own consulting or design firm. I have discussed on how to become an entrepreneur in Chapter-8 of this guide. Please refer this chapter for further guidance on things required and things to do before starting an entrepreneurship. **"But do not stop learning!"**

"When you do what you love. It's not work anymore."

J Balvin

3.1.2 Conclusion

In conclusion, I would like to say that it is important that you keep yourself abreast by following magazines and forums and keep yourself up to date with the recent trends in structural engineering. It is but usual that doing the same job day in and day out may make you a feel monotonous, but by following the magazines, forums and attending conferences, will make you love your job. And when you start loving what you do, you can propel yourself and fight with any setback which you may face in your career.

CHAPTER-4

Software Used in Structural Design

"When we started off we didn't know how to spell software."

Steve Jobs

4.1 Introduction

Manual structural calculations and analysis is a thing of the past, and structural analysis software are a way of life in the current structural engineering field. Without their knowledge, you cannot start a career in structural engineering as the days of manual calculations are over. There has been a lot of development in the field of software for analysis and design of buildings and structures; and it will be necessary for you to learn these before starting a career in structural designing. Software are accurate and save a lot of time while doing analysis and design. They are the new norm.

Nevertheless, the belief that a software will think for you is wrong. There is a tendency for structural engineers to learn the software thinking it to be the ultimate job getter. This is the kind of whimsical thinking that should be immediately removed. You should know that any software used in the design industry is a tool to assist you and save your time, it is not meant to work on behalf of you. The software delivers on what and how you input as a problem. Using of software also has its own standards, and there are accepted and good practices for use of any software, and these good practices can be downloaded from the software websites and can also be learnt on the go. Remember that the progression of finalizing a design goes through a certain process, and your structural calculations will be needed to be examined by a senior designer who will be acting as a checker; and the inputs you give and the results you get have to be disclosed to them first. That is the reason why a certain standard has to be maintained so that it is transparent and easy to understand, and errors can be avoided, and if occurred, can be easily recognized and removed.

Some of the coaching institutes who provide training for such software sometimes also lure students by providing 100% placement guarantees. Normally, such institutes provide placements for software operators and not for engineers. Your decision of joining an organization should be based on the potential the company provides for your development and training. Although, if you are hard-pressed in terms of your finances and are in need of a job, then you may choose to join just any company, that may be okay temporarily. But always be on the lookout for opportunities which give a wider scope of development rather than getting trapped in a situation where it would be difficult for you to wriggle out of later.

4.1.1 List of Software Commonly Being Used for Structural Design and Analysis

A number of reputed software are used by structural engineers to solve problems and design structures. Some of the renowned software are listed below and I can tell you that these are very user friendly and easy to use. You should try to be very conversant with them and learn how to use them for your analysis and design.

STAAD.Pro

STAAD.Pro is a 3D structural analysis and design software designed by Bentley Systems. It is very widely used and is acceptable in the community of structural engineering. It is also considered to be the most complete structural engineering software that can design and analyse almost every type of structure. It allows the design of structures in any material such as steel, concrete, timber, aluminium, cold-formed steel structures and for a wide range of loading conditions. In fact, it is a part of the syllabus of most of the engineering colleges and there is a good chance that you would be having a good knowledge of this software beforehand. Nevertheless, it is useful if you practice this software and have command on it.

STAAD.Pro releases its *infrastructure yearly handbook* and organizes the *year infrastructure awards program,* which recognizes the world's most outstanding infrastructure projects. The projects are submitted by Bentley System's software users and judged by a jury of infrastructure experts who adhere to the highest standards in determining which of the projects exemplify innovation, superior vision, and an unwavering commitment to exceptional quality and productivity. Bentley Systems shares the innovation in best practices among the infrastructure professionals who make these projects possible by featuring their outstanding achievements across the globe in 'The Year in Infrastructure' series of publications and special editions. The infrastructure series can be viewed free here https://www.bentley.com/en/infrastructure-yearbook. It also has a LinkedIn page with engineering insights for structural engineers which can be accessed here https://www.linkedin.com/pulse/staad-focus-project-diversity-joshua-taylor/.

Disadvantages: Although STAAD.Pro is a very versatile and accurate software you may need to be careful as it may give some uneconomical results for high rise buildings. Also curved or parabolic beam analysis is difficult to carry out in STAAD.Pro.

ETABS

Is an integrated building design software created by Computers and Structures Inc. Also known as CSI, it is a powerful software used for design of buildings and offers a set of tools for design of multi-storied buildings, from simple to complex structures, from a single-story structure to a commercial skyscraper. From the start of design conception through the production of schematic drawings, ETABS integrates every aspect of the engineering design process. Creation of models has never been easier - intuitive drawing commands allow for the rapid generation of floor and elevation framing. CAD drawings can be converted directly into ETABS models or used as templates onto which ETABS objects may be overlaid. The state-of-the-art SAPFire 64-bit solver allows extremely large and complex models to be rapidly analysed, and supports nonlinear modelling techniques such as construction sequencing and time effects (e.g., creep and shrinkage). ETABS provides an unequalled suite of tools for structural engineers which help in designing buildings, whether they are working on one-story industrial structures or the tallest commercial high-rises. ETABS has been immensely capable and easy-to-use since its introduction decades ago, and its latest release continues that tradition by providing engineers with the technologically-advanced, yet intuitive, software they require to be most productive. It is also especially useful in seismic applications. It is a software you should try to learn as it will teach you a lot about design of buildings and skyscrapers. Many companies who employ structural engineers train their employees in this software and this training is in their program for development of employees. However there are institutions which provide training for this program, you can access these at https://crbtech.in/. ETABS courses can be taken from Udemy also at cheap rates but you may need to have ETABS installed on your computer.

SAFE

STAAD.Pro and ETABS are used for designing the superstructure of a project whereas the foundations are generally designed using other software, and SAFE comes in handy while designing complex foundations and slabs. SAFE is also developed by CSI. It is an easy-to-use software for designers and can be imported or exported to CAD programs and excel spreadsheets. SAFE is a tool for designing concrete floors and foundation systems. From framing layouts to detailed drawing production, SAFE integrates every aspect of the engineering design process in one easy and intuitive environment. It provides good productivity to the engineer with its unique combination of power, comprehensive capabilities, and ease-of-use. Laying out models is quick and efficient with the sophisticated drawing tools. It also has a lot of import options that can be used to bring in data from CAD, spreadsheet, or database programs. Slabs or foundations can be of any shape, and can include edges shaped with circular and spline curves. You can access more of SAFE features on https://www.csiamerica.com/products/safe/features.

SAP2000

This is a design software for all engineering and design purposes and can be used to model and design many types of structural elements. From its 3D object-based graphical modelling environment to the wide variety of analysis and design options completely integrated across one powerful user interface, SAP2000 has proven to be the most integrated, productive and practical general purpose structural programs in the market today. This intuitive interface allows you to create structural models rapidly and intuitively without long learning curve delays. It is also produced by CSI along with ETABS and SAFE. You can find more details of the software on https://wiki.csiamerica.com/display/sap2000/Home. You can also find tutorials offered by CSI on https://wiki.csiamerica.com/display/tutorials/SAP2000.

PROKON

PROKON Structural Analysis and Design, is a suite of over forty structural analysis, design and detailing programs. The first PROKON programs were developed in 1989, and today PROKON is used worldwide in over eighty countries. The suite is modular in nature, but its true power lies in the tight integration between analysis, design, and detailing programs. Further details about PROKON can be accessed at https://www.prokon.com/.

REVIT

Revit is a Building Information Modeling (BIM) software widely used by architects, structural engineers. It is also used extensively in mechanical plumbing and electrical services. The software allows users to design a building and structure, and its components in 3D, annotate the model with 2D drafting elements, and access building information from the building model's database. Revit is a 4D building information modeling capable with tools to plan and track various stages in the building's lifecycle from concept to construction and later maintenance and/or demolition. It helps architects, designers, builders, and construction professionals to work together. The software is a sophisticated way to create models of real-world buildings and structures. Revit can be accessed at https://www.tekla.com/products/tekla-structures.

TECKLA Structures

Tekla Structures is used in the construction industry for steel and concrete detailing, precast and cast in-situ. The software enables users to create and manage 3D structural models in concrete or steel and guides them through the process from concept to fabrication. The process of shop drawing creation is automated. It is available in different configurations and localized environments. With Tekla Structures, you can create accurate, information-rich 3D models that have all the structural data you need to build and maintain any type of structure. Tekla Structures is known to support large models with multiple simultaneous users, but is regarded as relatively expensive, complex to learn and difficult to fully utilize. It is used for BIM and is a competitor of REVIT and can be accessed through https://www.tekla.com/products/tekla-structures.

4.1.2 Conclusion

There are a number of software available, and the most popular ones are listed in this chapter. It is always good to have knowledge of these software and the good practices used for operating these software. Nevertheless, most of the good companies which recruit structural engineers provide special training for engineers on how to use these software, and you will learn these easily as they are very user friendly and easy to operate.

CHAPTER-5

Professional Institutions and Associations

"Professional is not a label you give yourself - it's a description you hope others will apply to you."

David Maister

5.1.0 Introduction

Having a successful career requires more than a professional degree. Your capability, your competency, and understanding of the industry are equally important and professional institutions provide you this extra requirement. This chapter discusses about the various professional institutions which provide support to structural engineers and how you can benefit from these. Although it may not be tempting at all to join a professional institution or organization, during or after your battle with your studies; after all, who may have the time to attend meetings and conferences or engage in discussions, but thinking in such a way may cause you to miss out on the various benefits which a membership with a professional institution or association can get you. The membership of these institutions can make quite a difference in the build-up to your career as you can make a lot of professional contacts and get a huge wealth of information from these institutions.

5.1.1 What Are Professional Institutions or Organizations?

There are a number of institutions and organizations which label themselves as professional organizations. These are charitable non-profit organizations or bodies that seek to further the cause of a specific profession and their members. These are usually members of regulatory bodies who control the ethics and set the standards of an institution, association or an organization.

5.1.2 Benefits of Joining a Professional Institution or Body. There are a lot of benefits in joining a professional body as

1. **Extra Initiative:** Joining a student body shows that the student is taking extra initiative to learn more about their fields of interest.

2. **Networking and Mentoring:** You can deepen your ties with an organization by joining it, and in doing so, you will be increasing your network of contacts and can mix and get guidance and mentoring from some senior members of the body.

3. **Professional Development:** In structural engineering, training and development of your skills is a requirement which you cannot miss out on. These institutions provide you with regular

courses, either free of charge, or at subsidized rates so that you can master your skills.

4. **Conferences:** Students and members can attend conferences and engage with learned and experienced speakers. This exposure makes students familiar with the trends of the industry and they can impress their would-be employers with this.

5. **Professional Recognition:** Most of the institutions have an assessment process that applicants must pass before they are granted a certain grade of membership. Once you have achieved this grade, it indicates that you have achieved a certain industry-accepted professional level and expertise, and this adds to your professional credibility. But remember just being a card carrying member of an organization doesn't mean that this will automatically produce results for you –you will also need to analyse the different types of professional bodies or institutes and chose carefully. Simply collecting post nominal accreditation when taken to extremes can work counter-productively too.

6. **Career Development:** Continuous professional development (CPD) is essential if you want to forge a successful career after leaving college. Professional Institutions offer career development programs, training courses, and assessments. There may also be the opportunity to upgrade your membership to 'Chartered' or 'Fellow' grade via further

assessments, which gives added professional credibility.

7. **Industry Standards and Best Practices:** Most of these professional bodies advocate and teach the use of the best practices in structural designing. Using these practices in your design, show that your design is a professionally developed design which in turn will give you wide recognition in your field, especially amongst other designers and checkers. Since these standard practices keep on updating by bringing in new practices and discarding the obsolete ones, you will need to keep yourself abreast of the changes in this area.

8. **The Cost:** Professional bodies will charge for assessment and accreditation, and there will be an annual subscription fee, which could be in the range of £150 to £250. However, you may be able to claim professional membership costs against tax in certain circumstances. Also, investing in such professional institutions will give you an acceptance of being a member of the structural engineering fraternity.

5.1.3 Grades of Membership

There are a number of professional institutions that serve the field of structural engineering and offer different grades of membership. There are institutions which are renowned and have served the field of structural and civil engineering for nearly a century. These institutions offer different grades of membership and most of the common grades of membership which these institutions offer are listed below.

1. **Student Membership:** Nearly all the institutions encourage students to join as a student member. This grade of membership is free and gives a lot of benefit to a student, as they can gain access to lot of professional information and can start networking with professionals at an early stage. The membership grade on a CV shows that you have done something extra to gain knowledge and are ready to learn and develop your skills.

2. **Member:** This grade is for professionals who aspire to be a part of the organization or institution. For achieving this grade the professional institutions/organization require that you demonstrate your knowledge and skills. While some are strict in their requirements which can be exams or interviews which you may have to take, there are others that may give you a membership based on your education. Please check the requirements and the process to get a membership, of each professional institution/organization.

3. **Chartered Members:** This grade is a professional grade of membership and almost always requires you to demonstrate your competence by the way of examinations, case studies, and interviews. This is a grade which carries a lot of weight; and if you gain the chartered membership of a reputed professional body, it showcases your professional competency and the skills that you have acquired, and is an automatic recognition of these. Chartered membership of reputed professional

institutions like The Institution of Structural Engineers also known as IStructE, are a mark of professionalism and are respected worldwide.

4. **Fellowship:** Fellowship is the most senior grade of membership and is given to senior people in the industry who have gained experience and knowledge and have achieved a high level of seniority in their careers.

5.1.4 How to Make Most of Your Membership

You can always take advantage of being a student member or a member of a professional body as it bolsters your studies, your social life and job hunt and also your future prospects. For example, you can make contact with potential employers at networking events, and as I also earlier pointed out, these bodies host various events which will complement and broaden your knowledge. You can have access to their magazines and journals, and some of them even have libraries that you can utilize to enhance your knowledge. These memberships will give you plenty of opportunities. However, the only important thing is that you should engage yourself in their activities and always try to be a part of their events and conferences. Just by merely joining and not engaging with them will give you no benefit.

5.1.5 List of Professional Institutions

I have listed some very prominent and renowned professional bodies, institutions and organizations around the world who serve the field of structural engineering. These consist of professional communities which can mentor and guide you throughout your career and increase your knowledge, competence and professional skills. You can go through the websites of these institutions and contact them for any further information. I have provided the link to their websites which you can use.

Institution of Structural Engineers – IstructE

This Institution boasts of being the largest membership organization dedicated to structural engineering. It upholds standards, shares knowledge, promotes structural engineering and provides a platform to give voice to the profession. The institutions impose a high code of conduct which is an important factor in the professional career of a structural engineer. The institution was originally founded in 1908 as the Concrete Institute and later in 1922 came to be known as the Institution of Structural Engineers. IStructE also has a research fund which they issue with an aim to connect better with the industry. It offers three grants as of now and each is connected with research themes developed in partnership with the industry:

1. Undergraduate Research Grant.
2. MSc Research Grant.
3. Research Award

The Young Researchers' Conference of the IStructE showcases the most innovative projects in universities around the world and issues prizes for outstanding projects and achievements in public speaking.

IStructE provides different grades of membership as follows:

4. Student Member: Students already pursuing their engineering studies can opt for a student membership. A student membership of this institution will further pave your way to a graduate or chartered member. Student membership is free and you can move on to become a graduate member once you complete your education.

5. Graduate Member: You can become a member of the institution by applying to the institution on its website. Membership of IStructE is recognized throughout the world and is a mark of a highly trained professional.

6. Chartered Member: To achieve this grade you will need to go through training programs and pass exams of the IStructE, and their interviews. If you possess a Master's degree in Structural Engineering from an accredited university or equivalent, you can directly join as a chartered member. For more details, you can contact the institution's Qualification Panel which will guide you to the most suitable method of membership. A chartered membership of the institution is recognized throughout the world as a mark of the highest standard of examined professional attainment. Chartered members are given the chance to play lead roles in the design team as they can solve complex problems and demonstrate team management and leadership.

7. Fellow: Fellowship is the senior grade of membership. Fellows have fully developed careers, with several years of experience and generally hold responsibilities at a high level. Fellowship of this institution recognizes the excellence in structural engineering achievement.

IStructE supports members in India and other countries throughout the world. It offers membership grades form student to Fellow with different benefits. You can further explore IStructE on their website https://www.istructe.org/, and for India you can explore https://www.istructe.org/get-involved/regional-groups/asia/india/.

Structural Engineering Institute of ASCE (SEI)

The Structural Engineering Institute is a specialty institute of the American Society of Civil Engineers (ASCE). It is a world-renowned and respected institution, which is a professional body founded in 1852 and represents members of the civil engineering professionals worldwide. It has branches all over the world including in India. Through the expertise of its active members, ASCE is a leading provider of technical and professional conferences and continuing education. Their purpose is very inspiring and it reads as *"Help you matter more and enable you to make a bigger difference."* This Institute offers courses through its online platforms which you can access easily and study. The courses offered are of high caliber and of the highest professional standards. You can access the ASCE website at https://www.asce.org/structural-engineering/structural-engineering-institute/. You need to go through this website and explore the various membership grades it offers. It also has an Indian Section, ASCE-India, which was established in 1998, and currently has nearly 8000 members as of 2020. The annual fees of being a member in ASCE is £245, and is affordable considering the support offered by this society.

1. **Student Membership: The ASCE membership is free for students** and it has student chapters in major cities of the world. The Student Membership provides a lot of benefits in the form of conferences, education resources, and free webinars. You can access the Student Chapter page of ASCE at https://www.asce.org/student-chapters/.

2. **Membership:** Becoming a member of SEI (ASCE) is a mark of professionalism and can project you as an engineer who is familiar with the recent trends and is keeping up with the technological changes. You can explore further about their membership grades on https://www.asce.org/structural-engineering/sei-membership/.

3. **ASCE Library:** The ASCE has a vast online library from where you can access a lot of educational material. It contains the latest technical information available.

National Council of Structural Engineers Association (NCSEA)

This is a professional association in the United States of America, with member organizations of 44 states. The NCSEA was established in 1993. It was formed to constantly improve the standard level of practice of the structural engineering profession, and to provide an identifiable resource for those needing communication with the profession. The Association's vision is to be recognized as the leading advocate for the practice of structural engineering via its ongoing mission of representing and strengthening its 44 member organizations. The NCSEA serves the needs of the structural engineering profession, and its clientele. The association has a Young Members Group for young or new structural engineers, which also has a LinkedIn account https://www.linkedin.com/groups/5117836/. You can request to join this group directly through this link. This group provides a community for young or new structural engineers to interact with their peers while also eliminating the gap between schooling and licensed practice. Further details of the **Young Member Group** and licensing is available on the following link. http://www.ncsea.com/members/younggroups/

Engineering Council

This is a regulatory body for the UK engineering profession and sets and maintains internationally recognized standards of professional competence and commitment.

International Association for Bridge and Structural Engineering (IABSE)

The IABSE is a non-profit organization with a mission to promote and advance the practice of structural engineering worldwide in the service of the profession and society. It is a scientific/technical association comprising members in about 100 countries. It was founded in 1929 and has its head office in Zurich, Switzerland. You can join as a Young Member of this association. The IABSE provides a lot of e-learning options too, which you can take to enhance your knowledge. The membership of this institution will give you a lot of exposure to conferences and other publications of the Association. You can visit their website on https://iabse.org/.

The Institution of Civil Engineers (ICE)

The ICE is an independent professional association for civil engineers and a charitable body in the United Kingdom. Based in London, the ICE has over 93,000 members, of whom three-quarters are located in the UK, while the rest are located in more than 150 other countries. The ICE aims to support the civil engineering profession by offering professional qualifications, promoting education, maintaining professional ethics, and liaising with industry, academia and the respective governments. Under its commercial arm, it delivers training, recruitment, as well as publishing and contract services. As a professional body, the ICE aims to support and promote professional learning (both for students and existing practitioners), managing professional ethics and safeguarding the status of engineers, and representing the interests of the profession in dealings with the government, etc. It sets standards for membership of the body; works with the industry and academia to progress engineering standards and advises on education and training curricula. The ICE is a licensed body of the Engineering Council and can award the Chartered Engineer (CEng), Incorporated Engineer (IEng) and Engineering Technician (EngTech) professional qualifications. Members who are Chartered Engineers can use the protected title Chartered Civil

Engineer.

The ICE also offers free membership for students which can be further upgraded to a paid membership after the studentship is completed. It also offers a number of training and certification programs. The ICE can be reached at https://www.ice.org.uk/.

Institution of Engineers (India) [IEI]

IEI is the National organization of the engineers in India. It is one of the oldest organizations in the world and was established in 1920 in Kolkata, and was awarded the Royal Charter in 1935. The Institution of Engineers has more than one million members in 15 engineering disciplines in 125 centers or chapters in India and overseas. It is the world's largest multidisciplinary engineering society in the engineering and technology world and provides different grades of membership to engineers. Through a membership of IEI, you will have access to their academic material and seminars. You can access the website of IEC at https://www.ieindia.org/webui/iei-Home.aspx.

5.1.6 Conclusion

In conclusion, I would like to say that joining a professional body will undoubtedly give you the most needed exposure to the industry and how it operates. It will teach you the professionalism you need to practice in structural designing and moreover, give you the confidence and knowledge of the international arena.

Note: You may be able to find a lot of institutions claiming to be professional institutions or non-profit organizations set up for the purpose of advancing the knowledge of structural engineering. Such organizations will offer you membership at very low rates but remember that these institutions may not be reputed at all, and spending money on these will not help you gain anything professionally.

CHAPTER-6

The Importance of Training and Development and List of Reputed Institutes Which Provide These

"Learn as if you are not reaching your goal and as though you were scared of missing it."

Confucius

6.1 Introduction

This chapter precisely tells you that there is a lot of technical stuff which you still need to learn after you complete your engineering degree. I never liked studying after completing my final exams especially after I got my graduation. I just kept my studies aside and busied myself in looking for a job. It was only then that I realized the importance of training. You can get some sort of work after your studies; after all, that is what we all aim for and why we study engineering; but, remember it is important to start off your career professionally and be on a path of constant growth. When I completed my graduation, I received very little guidance; and so, I have tried to guide you on why should you take up training programs and invest in them and why these are necessary for starting your career and for your career growth. In this chapter, I have tried to lay down the advantages of these training programs and also have tried to list some of the renowned institutions which offer training programs. The list provided is only for your reference and does not aim to recommend any specific institute. It is also advised that you explore all the institutes mentioned in this chapter, and do your research on Google as well as consult and take advice from your mentors and peers.

The Advantages of Training and Development

"To know what you know and what you do not know, that is the true knowledge."

Confucius

Achieving an engineering degree may just not be enough to land you a good structural engineering job. Of course, it is important for you to have sound grasp of the theory which underlies any profession, but there are some technical skills you will need to learn through some specific and specialized training programs which will make you grasp some of the real-world challenges. Apart from these, you will get the chance to be taught and mentored by experts. I have listed down some other very prominent benefits of training below.

1. Training programs are meant to train you in new and specific skills and aim to make you job-ready. These can be taken at any time before or during your career and on line or by attending in person. Training programs cover a wide area and you can

explore and take a program in the area which you require to learn and elevate your proficiency in. These programs will teach you how to apply your knowledge in an area of your choice, and are designed to increase your performance and productivity.

2. By taking up training programs, you can kick start and improve your career prospects by showing your employer that you have the requisite skills. These are proof of your competence and knowledge and show your employers that you can successfully transfer your skills and knowledge into the workplace.

3. You will be job-ready and can finish the specific training offered by the employer earlier than the others, and will be able to start your job as you will have become more confident and able to adapt better than others who have not gone through these trainings.

4. It gives you a good advantage above the other candidates while applying for a job as it is more likely that your keenness and passion for doing something extra will be noticed.

5. The training programs documented in your CV will help you stand out amongst other candidates as these demonstrate and illustrate that you are more orgainsed and equipped compared to others.

6.1.2 Online Training, Development and Certificates

As stated above online courses are a great way to enhance your capabilities and your knowledge in the field of structural engineering. Many types of training and development courses are provided by many institutions which are recognized and respected throughout the world. I am listing some of these below with their website links so that it is easy for you to explore. Some of these programs will also make you a certified engineer which is a mark of trust and professionalism.

Structural Engineering Institute of ASCE (SEI)

The SEI provides a number of training programs which you can access through their website. The catalogue of training programs provided by SEI can be accessed through this link, http://mylearning.asce.org/diweb/catalog?_ga=2.17944634.2091820 675.1594093110-1154484041.1593922259. The training programs of SEI are very highly regarded and they give you the extra expertise through which you can enhance your career in any specific area. You can get a generous concession on these training programs if you are a member. These range from £250 to £450 or more. You can contact the institute in case you need funding and they may be ready to oblige if you present a strong case to them.

Institute of Structural Engineering (IStructE)

As discussed in my earlier chapters, IStructE is an institution committed for the promotion of structural engineering. It provides specific training courses to structural engineers which are the best of their kind and by taking these you can develop and learn. The courses can be accessed through this link: https://www.istructe.org/training-and-development/courses/.

Synergie Training

This is one of the leading training companies in the UK and is recognized throughout the civil engineering field. They provide training courses in a number of civil engineering fields and are the leading company for providing training for CDM Principal Designer. You can access their webpage for design-related courses through this link: https://www.synergietraining.co.uk/training/construction-infrastructure-training/.

ASTS Global Education

ASTS is a Singapore-based ISO 9001:2008 certified company which provides competent education training courses in various areas, and one of them is structural engineering. It has offices in most of India's metro cities and provides top-notch training in steel structures design. Its Steel Structural Engineering Design is a job-oriented program, and you can benefit from this program if you are interested in pursuing your career in steel structure design and detailing. It also provides a certificate for this course in association with NACTET (National Council for Technology and Training, India). You can view their website at http://www.astsglobal.com/index.php.

Pertecnica

This institute boasts of offering job-oriented training to engineers. Although it is rated as a good institute for project management, it also offers training courses in structural design. For more information, you can access https://www.pertecnica.in/construction/.

Dimensional Academy of Engineering

DA is a professional training academy with its head office in Mumbai (Andheri). It has branch offices in many major cities and offers courses in structural engineering. These courses are job-oriented and can be further explored at http://www.dimensionalacademy.com/.

Udemy

You may be surprised to note this, but Udemy is a platform where you can find very good training courses at very cheap rates. These are subject-specific and to a degree, quite professional as well. Although you may not get that extra benefit from the name of Udemy as compared to courses offered by SEI and other reputed institutions, but you can still gain a lot of good and professional knowledge from the tutors on Udemy.

For becoming a successful structural engineer, you should be open to educating and improving yourself by constantly seeking to acquire the most updated information. This is possible by engaging in webinars and conferences. The ideas given above in this chapter can guide you to build a successful engineering career, but ultimately your career is in your hands and this guide can only increase your knowledge of the things happening in practice; and in order to succeed you will need to work hard and in the right direction, learn and keep on developing yourself, and that is one thing which no one can do on your behalf.

6.1.3 Other Useful Resources for a Structural Engineer

I have listed some of the free resources like forums and blogs which you can join as a structural engineer. Some of them may require you to donate some amount for maintenance of their blogs and forums. However, these are great ways of getting to know what is happening in the field of structural engineering.

Magazines

Structure Magazine: https://www.structuremag.org/

Civil+Structural Engineer Magazine: https://csengineermag.com/

Structure Magazine NCSEA:
http://www.ncsea.com/publications/structuremagazine/

The Engineer: https://www.theengineer.co.uk/

Civil Engineer: https://www.thecivilengineer.org/

Blogs and Forums

Structural Engineering Forum of India (SEFINDIA)

This is an independent forum for structural engineers, which requires no previous membership or affiliation. You can create an account and start interacting through discussion groups, articles, and forums for queries. This is a very vibrant structural engineering forum, and I strongly advise you to visit this forum. You can also donate a minimal amount and become a member. This forum has many professionals as active members who give solutions for day-to-day issues a structural engineer faces. The forum has a rich collection of articles and useful material on all aspects of structural design. It has 24747 registered users as per their records and 76785 articles have been posted by the users of this forum till date. Their website is https://www.sefindia.org/.

The Structural World

This blog has a lot of information on structural engineering and the usage of different types of software in this field. It can be accessed at https://www.thestructuralworld.com/topics/.

Dennis Mercado's Structural Engineering Blog

This blog has a lot of information available on CSI software. Have a look at https://dennismercadosstructuralengineersblog.wordpress.com/.

CSI Knowledge Base

The CSI Knowledge Base is a searchable online encyclopedia that provides information to the structural engineering community. Its purpose is to assist the users of the CSI software application and enhance their understanding within the field. Discussions on the blog are of general structural topics. Visit the blog at https://wiki.csiamerica.com/display/kb/Home.

iMechanica

This is a forum on the subject of mechanics and can be useful for structural engineers and can be accessed through https://imechanica.org/.

ENG-TIPS.Com

This is a very extensive forum for engineers and has a huge number of members. You can join this community and benefit from the discussions of the members. Visit the forum at https://www.eng-tips.com/.

Engineer Boards

This forum contains a lot of discussion on different topics of structural engineering. Visit the forum through http://engineerboards.com/.

6.1.4 Conclusion

Training, as mentioned above, is a great way of preparing yourself with the industry practices and making you ready for your responsibilities and brace any challenges that you are faced with. They look great on your CV and are a sure shot way of attracting any company HR's attention. On the other hand, magazines and forums are a great way to be abreast with all the recent trends and changes happening in the industry and also help in increasing your general knowledge in the construction industry. Joining forums will help you confront the problems you may face in your day-to-day work. You will be surprised by the amount of support you get through these forums.

CHAPTER-7

Some Reputed Companies and the Job Functions of a Structural Engineer at Various Levels

"You have to play for your Position."

Anthony Rougier

7.1 Introduction

This chapter will give you an overview of areas where you can start your career. It will give you an opportunity to reflect upon the skills and qualities you possess which you can utilize to decide which area to start your career in. In this chapter, I have also tried to describe some of the renowned firms with their websites so that it is easy for you to explore these companies in detail.

7.1.2 Where to Start Your Career

You are a structural engineer and being a structural engineer is in itself a very big achievement and you should be proud of that. Now you will be seeking an employment opportunity and making up your mind on where to start. Try to join big and reputed firms as a trainee or as an apprentice. This will give you the opportunity to train and learn and also will inculcate professionalism and know-how of the industry in you. You can start your career in small and midlevel firms but do not be content with those, always strive to grow. It is unfortunate that sometimes a structural engineer starts taking up petty work and gets involved and too much entangled in these, that he loses his desire to learn further. Such type of jobs may benefit you for some time, but eventually, you will hit a plateau, a blockade in your career and your growth may just stop. You may not be able to grow further in your career. This situation has to be avoided. There is no harm in structural engineers taking up small jobs for local consultants, but you should always strive to learn and be on a continued path of development of your professional capabilities. The first thing you need to do now is to make up your mind and decide where you would like to work. There are various options available for you some of which I have listed below.

Work With Contracting Companies

Working with contracting companies will be a great experience for you. You can get to work in a diverse range of projects and learn to optimize your designs. Contracting companies take up EPC (engineering procurement and construction) jobs and these require them to design and construct a project. A big contracting company takes up jobs in nearly every sector and you may get the opportunity to design various types of structures in your life-time. You will learn to design structures with an emphasis on saving cost without compromising on the safety of the structure. A major advantage of working with a construction company is that their remuneration is amongst the best in the industry. Some of the standard Fortune 500 contracting companies have standard structural design offices with very good facilities and very good chances of promotion and increments. These have standard work environments and proper hierarchy and structure in place.

However, the downside of working in a contracting company is that you may not get to research very much. As these companies tend to award complex jobs to experienced consultants, although this might not be the case with larger companies.

Work with Developers

Developers normally get their designs done by a consultant; nevertheless, some big developers may sometimes opt to hire structural designers to oversee the work of the consultants. This role is normally a supervisory role suitable for experienced engineers.

Work with Architects and Design Firms

Architects and Design firms typically work side by side with each other. In fact, some architects also hire designers for designing their structures. Getting a job with an architect firm will open up venues for you to work mostly in buildings. This work will teach the importance of coordination with other consultants like MEP, Fire Fighting, and landscape.

On the other hand, you can expect a lot of diverse type of work in a structural design firm. Good and standard design firms and departments engage in challenging and distinct projects. They have standard ways of working and a proper hierarchy in place. Although this may not always be true with small firms where you may have to do a lot of multitasking.

The remuneration in design firms is standard and may not be as good as contacting firms. Although international design firms have a class of their own and can offer attractive packages to a good and able designer.

Work with Government Undertakings

Working with government undertakings will involve some designing at the initial stages but as you progress, you can move on to the roll of an administrator, where the role requires administration skills rather than engineering skills. The main benefit of government jobs is that they are very stable and salary packages are standard with moreover standard raises on a regular basis.

Selection for these jobs is based on standard competitive examinations generally and are conducted by the government bodies made for this purpose. In India, you can sit for the Indian Engineering Services Exams if you would like to join the Central Govt. engineering services. You can explore for the various examinations and tests which the government agencies conduct in your states and countries. There are a number of undertakings which require structural engineers and will have their own systems for recruitment.

The downside for this is that structural design jobs in government firms are scarce and even if there is an opening it may take ages before you are finally called to join. Government firms or undertakings have their own standard policies which they cannot bypass. In contrast to this, private undertakings or companies in the private sector are more vigilant in recruitment and sometimes offer better packages and overall have a faster recruitment process.

Work as a Structural Consultant

I have detailed this out in Chapter-8; however, I would like to reiterate here that whatever be the case, it is better to start your own firm only after you have a good amount of experience and have learnt the trends and manners of the trade.

7.1.3 Some of the Renowned Companies and Firms who Offer good jobs and growth

I am listing down some renowned design firms and companies who hire structural engineers on a regular basis. This is where you should try to get a job as a junior designer first. Do not look for the money, instead look for the experience you will gain from working with them. Also, try not to choose small firms where you tend to get stuck up in a single type of work, but strive to learn and enter into different fields and get experience in all the areas. For example, if you join an architectural firm dealing with housing projects you may just end up designing some mid-level buildings and will not get exposure in multistory tall buildings or skyscrapers, and also ancillary structures. While you work, your focus should be to tread on a constant path of self-development. Keep in constant touch with your professional institutions while you work and engage in their activities too. Keep training, keep learning and keep gaining confidence and raise your professional competencies.

List and Description of Some Reputed International Firms around the world and in India.

ARUP

This is a multinational professional firm headquartered in London which provides architecture, design, planning, and structural designing services. The firm was founded by Sir Ove Arup in 1946 and boasts of 15,166 staff members worldwide in 104 disciplines. In a survey by the New Civil Engineering publication recently, 96% of the employees of Arup agreed that they were satisfied with the work culture. Arup is a well-established company with a large portfolio of construction and infrastructure projects across the world including Europe, Middle East, India and some other countries, nearing a total of 30 nations. It follows the highest ethics and offers excellent training and career progression and scores highly on pays and benefits.

You can register your interest as an apprentice, a graduate or in an internship with the company at https://careers.arup.com/earlycareers/. Your membership and training from a reputed institution will give you an edge above the others in this company. For details of membership and training see Chapter-5.

ATKINS

Atkins is a British multinational engineering, design, planning and architectural firm. It is a member of the SNC-Lavalin group of companies and has its headquarters in London, UK. It was founded in 1938 by Sir William Atkins and boasts of one the best job satisfaction among its employees. It has a global presence and has expanded into a wide range of sectors including aerospace and high-speed railways. It has an approximate 18,000 employees worldwide and is the third largest design firm in the world as per ENR 2019. Atkins provides openings for fresh structural engineers and provides training and professional development programs. It is a professional organization with a high rate of employee satisfaction and is the benchmark of the best industry practices in each trade. You can access its career page at https://careers.snclavalin.com/atkins-structural-engineering-jobs.

Stantec Inc

Stantec Inc is an international professional services company with offices all over the world including US, UK, India, and the Middle East. It provides professional consulting services in various fields of engineering and architecture. The company provides services on projects around the world through over 22,000 employees operating out of offices across 6 continents internationally. It provides programs for continuous learning and development to its employees so that their career keeps developing at all levels. If you are a fresh graduate or an experienced structural engineer, you can send your CV and credentials to them on their website and show your interest in joining their team.

Most certainly if you possess the correct skills and training that are up to their standards of expectation, you will get a call from them. You can access their career page here https://www.stantec.com/en/careers.

Thornton Tomasetti

This is a firm which specializes in structural design amongst other fields, and operates from more than 50 countries worldwide with about 1500 employees. The firm has provided structural design for the famous Petronas Towers in Kuala Lumpur, Malaysia and the Taipei 101 in Taiwan. It has affiliation of most of the renowned societies like the ASCE, the National Academy of Engineers, and the National Council of Structural Engineers Association. Thornton Tomasetti recruits young engineers and fosters and develops them. You can visit their career website and search for jobs or just drop your CV for future considerations. Their career page can be accessed through the following link. https://www.thorntontomasetti.com/careers.

MECON Limited

Formerly known as Metallurgical & Engineering Consultants (India) Limited, this is a public sector undertaking under the Ministry of Steel of the Government of India. It began in 1959 as the Central Engineering and Designing Bureau (CEDB) of the Hindustan Steel Limited (HSL), the first public sector steel company. Subsequently, CEDB grew as Metallurgical and Engineering Consultant (MECON), a subsidiary of the Steel Authority of India Limited (SAIL) in 1973. It later became an autonomous company reporting to the Ministry of Steel in 1978. 1st April is observed as the Foundation Day in MECON. Mecon Limited is one of the most professional Government of India undertakings, and offers good professional careers in structural designing. It caters to various areas of civil engineering like oil and gas, power, steel industry, and infrastructure. It has a career portal with the recruitment process which can be accessed through http://www.meconlimited.co.in/Recruitment_process.aspx.

L&T Infra Engineering

L&T Infrastructure Engineering Ltd. is one of India's leading engineering consulting firms offering superior technical services in transport infrastructure. The company has extensive experience both in India and globally, delivering single point 'Concept to Commissioning' consulting services for infrastructure projects like airports, roads, bridges, ports and maritime structure including environment, transport planning and other related services. L&T Infra has offices in Chennai, Hyderabad, New Delhi and Mumbai. It is driven by high ethical values. It designs highways, bridges, airports, marine services and is a part of Larsen and Toubro. It continuously advertises for structural design vacancies and you can explore these on their career site at http://lntiel.com/index.php/careers/. It offers structured training for young designers, latest methodologies and tools, and adheres to best industry practices. You can also follow their LinkedIn page on https://www.linkedin.com/company/l&t-infrastructure-engineering-limited/about/ and keep yourself updated on the latest trends.

Afcons

Afcons is a part of the Shapoorji Pallonji (SP) Group which is the second-largest construction group in India. Afcons has emerged as a leader in Engineering Procurement and Construction Projects (EPC), delivering projects in the areas of marine, highways, bridges, metro, tunnels, etc. The company has presence in 12 countries including India, Africa and the Middle East. It has a LinkedIn page which you can access through this link: https://www.linkedin.com/company/afcons-infrastructure-limited/?originalSubdomain=in. It also has a career page on its website which can be accessed through the link http://careers.afcons.com/.

JaiPrakash Associates

Jaiprakash Associates Ltd is an Indian Conglomerate based in Noida, India. It is a diversified infrastructure company with business in Engineering and Construction, Power, Cement, and Real Estate, etc. It is a leader in construction of projects of river valleys, hydropower projects on a turnkey basis. The organization boasts of identifying every employee as an achiever who will make a difference. There is a strong emphasis on networking and knowledge sharing with others to ensure to work together as an organization. It also provides employees with good facilities. Its career page can be accessed through the company website at http://www.jalindia.com/index.html.

Hindustan Construction Company Ltd. (HCC)

HCC is a pioneer in India's infrastructure industry, having executed landmark projects that have defined the country's progress since 1926. HCC Ltd. is a public-private company with interests in Engineering and Construction, Real Estate, Infrastructure, and Urban Development & Management. The HCC Group of Companies comprises of HCC Ltd. and its subsidiaries HCC Real Estate, HCC Infrastructure Co Ltd. HCC is also developing Lavasa, a planned hill city set in the Sahayadri Mountains. Lavasa is spread amidst 23,000 acres and located at a 3-hour drive from Mumbai and an hour's drive from Pune. The master plan for Lavasa has been developed by the design consultant HOK International Limited, USA. Lavasa is slated to have 5 towns - Dasve, Mugaon, Dhamanohol, Sakhari-Wadavali and the Central Business District (CBD). The city is planned for a permanent population of 3 lakh residents. The integrated development at Lavasa City comprises apartments, retail, hotels, an international convention center, education, information technology, biotech parks, sports and recreation facilities spread across its 5 towns.

HCC encourages young engineers and graduates to join and has a special portal for young engineers and recent graduates which can be accessed at https://www.hccindia.com/career/recent-graduates.

HCC has constructed a culture that encourages people to venture beyond the brief and to think big. Their workplaces and project sites are the most sought-after training grounds for young engineers.

Gammon India

Gammon India is amongst the largest physical infrastructure construction companies in India. Its track record spans significant landmark projects built over several decades, with a prominent presence across all sectors of civil, design and construction. It has a track record of building landmark structures, some of which have become iconic. This includes 'The Gateway of India', the piling and civil foundation work for which was successfully executed by Gammon as its maiden project way back in 1919. Gammon's expertise also covers design construction and operations. Gammon International includes a majority holding in SAE Power Lines, and Sofinter Group, Italy spanning the sectors of power and industrial boilers as well as waste and environment management systems. Gammon has received accolades and recognitions from a variety of reputed institutions. Its website can be accessed at http://www.gammonindia.com/home/gammon-india.htm.

NCC Ltd.

NCC Ltd. undertakes civil construction in segments such as buildings, water, roads, irrigation, power, electrical, railways, metals, mining and has also a presence in the Middle East where it currently undertakes works in roads, buildings, and water segments. It is currently 129th in the top 500 companies in the Dun & Bradstreet publication. It has a separate structural design division which caters to the design requirement of the company. The career page of NCC Ltd. can be accessed at http://ncclimited.com/careers.html.

Punj Lloyd

Punj Lloyd is an International conglomerate offering EPC services in energy, infrastructure services along with engineering, and manufacturing services in the defense sector. From pipelines, tanks, terminals to refineries and airports; Punj Lloyd has developed many landmark projects for ONGC, IOC, Reliance, Shell, Qatar Petroleum, Petronas, ADNOC, and has emerged as a strong player in defense, pursuing various programs under the land systems, aerospace, etc. It boasts of about 8775 strong, skilled, and multicultural workforce. The group believes in nurturing talent and offers a commitment to enhance learning and nurturing young talent so that they can handle new projects independently. The career page of Punj Lloyd can be accessed through http://www.punjlloydgroup.com/careers/learning-development.

GMR Group

The GMR Group is a major player in the infrastructure sector, with world-class projects in India and abroad. The GMR Group is headquartered in New Delhi and has been developing projects in high growth areas such as Airports, Energy, Transportation and Urban Infrastructure. The GMR Group is one of the fastest-growing infrastructure enterprises in the country with a rich and diverse experience spanning three decades. With their vibrant portfolio of projects, GMR is uniquely placed to build state of the art projects in sectors that are of critical importance in the process of development. The organization endeavors to provide job opportunities across businesses, depending upon company needs and professional interest and competency of the employees. As an organization it offers employees internal growth opportunities – from Airports to Energy to Highways to Corporate and vice-versa, all under one umbrella. The career portal of GMR can be accessed at https://www.gmrgroup.in/careers/.

Note: I have not listed down some of the Indian Govt. undertakings for the reason that these have their own processes and may vary from State to State. The current trend is that government undertakings and entities outsource their work to private firms and job opportunities are quite less compared to the private and public undertakings. However, if you are interested, you can search for these companies like ONGC, NTPC, IOCL, etc.

7.1.5 Layout of Job Responsibilities in a Typical Design Office

Normally a structural designer works with construction companies such as listed above and also with developers and architects. There are separate design divisions in good companies with proper hierarchy in posts. I have given the hierarchy of positions in a standard design office that is followed internationally and in good consulting design offices in India too.

In such divisions, the usual spread of responsibilities will depend upon the size and the capacity of the office and the size of projects that are taken up for design.

Assistant Designer/ Junior Designer

An Assistant Designer's job is to design structural elements of a structure, like foundations, columns, etc. depending upon the size of a project. There may be a number of design assistants working together on a single project at the same time. At the initial stage in your career, you will be inducted as a Junior Designer or Assistant Designer. There are a number of names for this post depending upon the company, and this will be your first posting in most of the cases. You may be working under a Senior Designer who will be allocating you responsibilities and work and will be responsible for your overall performance. You will start to learn what designing means in this position.

Your early responsibilities will be to design sections or elements of a project and work as part of a design team headed by the Lead Designer.

Detailers

A Detailer is a draftsman working as a separate team and will usually be working under the Designers and the Assistant Designer and are responsible for producing and detailing the design. Detailers are a very important part of the team and are experienced in drafting structural designs using standard detailing practices and procedures. It may help you to learn the standard practices used in detailing at this time in your career.

Checkers

Checkers function as the checkpoint and are responsible for checking the design before they are released for construction. They are senior designers with good experience of structural design and are responsible to detect faults and errors in a design. They are also responsible to check the overall safety and whether the structure is as per the standards and complies with all the prescribed laws of the state, if any. You can advance to the role of a Checker once you gain experience.

Designers

Designers are responsible for single projects and may have several Assistant Designers and Detailers working under them. They will be leading a team and will be a mentor for the Assistant Designers. They are also responsible for coordination with architects and other MEP work.

Lead Designers

These are team leaders who lead all the above categories. This post is available in large design consultancies or offices and a Lead Designer will be responsible for the overall performance of the design team. There may be a single Lead Engineer or a number of them depending upon the size and requirement of the office. These will be licensed by the concerned statutory authority as is required in some countries, or chartered members of organizations like IStructE, SEI or NCSEA. They are responsible for approving the final designs which are then released for construction.

Chief Engineer

The Chief Engineer functions as the head of the whole office and is incharge of the whole process of design of all the projects. Usually, a Chief Engineer's post is a very senior post responsible for communicating goals of the firm to all employees, delegating tasks, negotiating with clients, and business development. A Chief Engineer is a highly experienced engineer and should be, in most cases, licensed as required by the statutory authorities and also can be a fellow of a reputed professional institution.

CHAPTER-8

Starting Your Firm

8.1 Introduction

Starting your own business is advisable only after you gain considerable experience in the areas in which you want to start. Remember that structural design concerns public safety at large and you should be well conversant and thorough in your work before you move on this expedition. This chapter will take you through the various requirements for starting your own firm and working independently as a structural designer. I will try to pack in the practical stuff and guide you on how you can plan and start your own firm. The points which I have highlighted have been written after going through relevant facts and substance, and after experiencing and seeing a number of designers flourish or fail.

8.1.2 Why to Start Your Own Business

The best thing about structural engineers is that they are a living and readymade working factory which can churn out marketable products at a decent rate without the need for heavy investment in machinery and other paraphernalia. A structural engineer can start their work with as little as themselves, their computer and a pair of software. Whatever initial funding is required is basically for the marketing. To begin with, you can think of building a startup from your own home and then move on to have a nice office in a niche area, as keeping costs down should be a prime objective at this stage. The advantage you have in this business is that you do not need a huge investment to start, and risks are low. At the most, you can return to take up a job in case of failure.

Most importantly, if you want to start a flourishing business, you need to be patient and work with resilience and be sturdy and steadfast in your pursuit of success.

8.1.3 Carve out a Niche for Yourself

There are several specialties within structural engineering, and having a niche in any of these will help you be perceived as an expert in that area. You can become an expert in designing towers, buildings, bridges, stadiums, or any other structure. The only factor that you need to note is - start your career in an area where you can get some work. Some areas catering to special fields may not have much regular work as a startup firm, like design of bridges, airports, etc. Hence, it is better to look for an area where construction takes place regularly like buildings.

8.1.4 Professional Membership

If you are trying to start a consulting firm, it is a good idea to apply for membership of yourself and/or your firm in any of the professional institutions, especially organizations like the Engineering Council, or the American Council for Engineering Companies or any other similar professional body. The membership of these bodies is a mark of quality and excellence in their fields and your firm will start to be recognized. Once you are a member of these bodies you will be following their ethics and good practices. This will be a mark of trust in your profession. Apart from this, once you are a member of any of these institutions, your clients will respect your firm and your firm can easily qualify in most of the reputed companies' tender processes.

8.1.5 Experience

If you want to establish a foothold as an independent organization, you should gain the required experience and expertise of the field you want to work in. Though there cannot be a defined number, a minimum of eight to ten years of experience should suffice. It is very hard to convince your clients if you do not have the required experience, as structural engineering is a field which requires you to be very diligent and careful in your work. After all, it is responsible for the safety of people who will be using your structure day in and day out.

You should also learn to collect the latest information on design trends. Your professional membership may help you here. You can also approach your professors, seniors or other structural engineers for their knowledge and experience. Usually, structural engineers like to share information and knowledge with each other, and you will be surprised to see the cooperation you can get at times.

8.1.6 Check the Statutory Requirements

Some countries have special statutory requirements which you need to fulfill before you can start designing independently. Countries like the USA have requirements for licensing of structural engineers, and this varies from state to state and can be explored from various sources. I will not delve on the specific requirements in this book. However, it would be a good idea if you can look through the Structural Engineering Licensure Coalition (SELC) stance on this matter. The SELC includes all major organizations representing engineers throughout the United States and is dedicated to a common position in support of structural engineering licensure nationwide. The SELC is comprised of Structural Engineering Institute (SEI), the National Council of Structural Engineers (NCSEA), the Structural Engineering Certification Board (SECB) and the council of American Coalition of American Structural Engineers (CASE). I have dealt with all these in my earlier chapters.

On the other hand, some countries may not require a license for practice as a structural engineer, and your degree or your Master's degree with some demonstrable experience will suffice. India is one country where you do not need a license to practice as a structural engineer and in most of the states your qualification as an engineer is enough for you to start a firm. Nevertheless, you may need to register with some or the other tax authorities like the Service Tax authority to get a Service Tax Registration Number.

8.1.7 Funding Your Business

You may be needing financial support to start your business, and operating it. Although I would recommend that you start small by, may be, starting an office from your house, working solo, and as a sole proprietor rather than immediately starting a corporation or a company. However, this all depends upon how you are placed and how many customers you already are dealing with. Governments are generally supportive of startups and you can apply for grants and funding from many of your Government schemes for startups which you can explore.

Financing options will depend upon the growth stage of the company, and at initial stages, it is a challenge to arrange finance as investors face a high risk at this stage. It will be wise for you to explore the grants which your government can provide and also see if you can self-finance with the help of family and friends in exchange of 5% to 10% profit sharing in your firm. This will depend upon how much support you can garner from family and friends. Crowd financing is another option which you can look into.

It is always wise to keep your cost low at the start up and you can do this by following some of the options set out below. Although these are not very lucrative, but they certainly can keep your costs down to comfortable levels.

- Sole Proprietor to Corporation: Start of as a sole proprietor and work from home as this will save you rent and office set up and operational costs. You can gradually move on to be a corporation and have a nice office. Keep your startup costs at the minimum.

- **Learn how to keep your accounts:** Become an accountant or hire one. But you may need to understand accounting and keep your accounts initially to avoid the accountant's cost and also to avoid any undue penalties from the concerned tax authorities.

- **Learn Marketing** and become your own salesman for the initial period. You may need to move around a bit meeting architects, developers and other potential customers. Keep on meeting them, your networking will help you here immensely.

- **Learn drafting** techniques and detailing and do your detailing yourself until you have enough work. You can hire freelance Detailers later on for drafting and detailing your designs. You can get these on an hourly basis. Choose one or two of them who you think are really good. Sometimes these Detailers can also act as your salesman if you build a good relation with them.

- **Learn to understand contracts** and write like a lawyer. You may need this skill to draft contracts and also specifications and notes which can avoid your liability.

8.1.8 Insurance Cover

Check the insurance requirements. You may need a Professional Indemnity Insurance for working on reputed projects. This is a necessary requirement and you should have one irrespective of where you work. This will protect you from any professional negligence in case you commit one. You may also need a General Insurance cover, once you start grow to a certain level, for your employees and for your office. Look into the various options and explore the various insurance policies. It may be wise to explore this area in advance as you can have a well-laid plan to deal with this cost.

8.1.9 Try to Have Some Clients in Hand before you Cut Loose

It is always advisable to network and have some clients that you are confident you can get work from, before starting a full-fledged firm. You can start your firm by yourself as mentioned earlier, first from your home instead of spending on an expensive office and other requirements like furniture, etc. It is better first to operate as a sole proprietor, and as your business grows, you can shift to a rented office. It does not matter from where you work as long as the work you are doing is professional, standard, and as per your client's liking.

8.2.0 Some Essentials Which You May Need to Know

You should be a master of many trades. You should be quite conversant in your area of design, such that the firms or clients you are working with, develop confidence in you. For example, if you are an expert on buildings, you should have the knowledge of the MEP (Mechanical Plumbing and Electrical) requirements of a building and need to be wary of these while designing. It is a fact that when you start your firm, there will be a general expectation amongst your clients and they would expect you to have knowledge of all the requirements in the trade; and hence, they may expect you to come up with new ideas and problem-solving skills. Thinking out of the box can matter a lot here, and structural engineers who are confident and have great ideas in their minds can make a big difference in the market. You should also have the knack of understanding your client first and then their requirements.

Consider the following facts

- **Your initial advertisement** will be through word of mouth until you grow into a large firm. But to grow into a large firm you will need loads of work! Please check and recheck your work before you authorize any designs. Any mistake can harm your reputation and take away your business.

- **Networking:** Have good networking within the engineering community as this will bring you business and act as a great advertisement platform.

- **Check your funding.** You should have enough funds and sources of funding so that you can sustain the initial expenditure of your firm for at least six to eight months, as you may not level off before this time.

- **Create a nice and professional website** and brochure. You can get this done easily through some of the freelance sites such as Fiverr or Upwork. Always keep these updated as out of date information will show that you are careless and may cost you some clients.

- **Always be available.** Always answer phone calls and emails or social network messages. Be positive and make it a point to connect if you have missed a call or a message.

8.2.1 Start looking out for Big Contracts and Tenders

Big corporations and government undertakings in the public and private sectors call for bids from architects and structural engineers. Once your firm is registered, you can start bidding for such tenders. Most of them are in the lookout for an experienced and skilled structural designer.

Initially, you may face some setbacks, but as you progress you can partner with some other bigger firms to win a bid and then the doors can open for you for further contracts. This may require some networking and some advertising techniques. With your low establishment costs, you can bid competitively. As per ContractorCalculator CEO, Dave Chaplin, contractors "need to acquire and apply marketing, sales, negotiation and business skills" to win work from clients. You may find it useful to look at the following skills which can help you in winning some good contracts.

- Have a good killer contractor CV which is impressive and to the point.

- Understand your client and provide good and practical solutions.

- You need to understand one thing that your client wants not only safe structures but also structures which are economical. So study the project and try to give the best solutions.

- Try to invest time and money in developing contracting skills as these will help you in winning more contracts and you will last longer in the business.

- Always remember to cover yourself with a Professional Indemnity Insurance when working on large projects.

8.2.2 Conclusion

In conclusion, I would like to say that starting a firm for a structural engineer can be as easy as purchasing a laptop and software, but you should not venture into this business unless you are well experienced and confident. As per recent surveys of the labor department and other major surveys, there will always be a demand for structural engineers as new structures come up and old structures are required to be refurbished and repaired.

CHAPTER-9

Some Tips on Writing Your CV

"Your CV says a lot about you, It determines whether you will be called for an interview or not."

Anonymous

9.1 Introduction

The CV is your marketing tool. Remember, you need to sharpen it and make it market ready. I will try to explain and give you some tips on how you can write and polish your CV.

CV is short for the Latin word "curriculum vitae" which means "course of life." It is a document which should highlight your achievements and your professional history. Your professional history can include your education, your professional memberships, your trainings, your IT knowledge, your professional work experience, etc.

9.1.1 Some General Tips at First

- It is always a good idea and is acceptable to have different CVs for different job requirements. That is, you can juggle between your skills and highlight the ones that are required for the job you are applying for. Remember, there is no standard format for a CV. It is a document which aims to project you as the worthiest candidate for the job opening you are applying for. So have a format ready in which you can change your skills and other details to fit the job requirement you are applying for.

- **Plan your CV**: Before you start writing your CV, plan it. Think of what you need to write, of what is relevant and what will attract the attention of the employers. Think how you are going to structure it. Try writing something down, as this will help you start writing your CV.

- **Writing your CV:** First start with the facts, your name, and address, date of birth, qualifications, and trainings. The sequence of education and experience can be juggled, but it is better to mention your education first and then write your experience. If you already have some experience, write only the relevant ones.

- **Presentation:** Your CV should have a proper presentation so that your strengths immediately pop out once it goes into the hands of your prospective employers. Think of yourself as a salesman marketing yourself. Check and think about the structure and the style that would be appealing as well as professional. You can also search for proper structure and style of your CV on the internet.

- **More on the Structure:** Your CV should have your name, your address, email, telephone number at the top of the CV. Then mention your skills, education, and experience. Each heading should be clear and visibly well presented. Examine your CV's look and check for formatting errors as well as the print layout and blank spaces.

- **Be Concise:** Do not let your CV run into more than two pages. Try to cut it down if it does.

- **Vocabulary:** Try not to repeat the same words or phrases multiple times. If you are typing down your CV in MS Word then make use of the Thesaurus, you will get a lot of synonyms that you can choose out of. A good use of vocabulary makes your CV professional and easy to read.

- Do not let your employment history have gaps, try to cover them with some useful details, such as trainings, etc. which you might have taken up during this period, or write down honestly what you did and what you learnt but make it relevant to the job you are applying for.

- Do not waste words and time on your objectives, nobody cares about them and nobody reads them and they do not matter really, instead spend some space on writing down some relevant skills that you have acquired.

Hundreds of articles on the rules of creating a CV have been written by recruiters, psychiatrists, career advisors and if you Google this, you will find about 106 million articles on the topic, which include hundreds and hundreds of advertisements; but, there are some points which you need to keep in your mind while writing your CV.

1. **The Purpose of Your CV:** Consider a CV as your sales pitch. It is there to progress you to the interview; it is not a detail of your life story. Be precise and highlight the areas on which you want your prospective employers attention.

2. **Be Aware Of the Hiring Manager:** Once your CV reaches in the hands of your hiring manager, they will hardly take more than six to ten seconds to skim through it post which it goes to the 'yes' basket or the 'no' basket. Make your CV visually appealing and compelling enough for the hiring manger to take you to the interview stage.

3. **Be aware of the Interviewer:** Once your CV passes the hurdle of the hiring manager, it goes in the hands of the interviewer who will use it as a tool to ask questions and try to learn more about you. Be honest, as you may have to answer a lot of questions regarding your skills and your knowledge which you have highlighted in the CV.

4. **Know exactly what the hiring manager** is looking for: Frame your CV in a fashion that your CV is recognized and picked up easily. A hiring manager may be looking for some special skills and requirements; try to highlight those which are relevant for the job that you are applying to.

5. **Be factual and be Humane:** Write something in your CV about

yourself in a line. This helps to show that you are humane and always document facts, as these will be picked up by the hiring managers who are generally smart and are paid to do exactly this.

9.1.2 The Covering Letter

It is always recommended to include a covering letter whenever you apply for a job. It is a one-page document which highlights why you consider yourself a candidate for the job and why you should be hired and goes beyond the CV to explain how you are going to add value to the employer. It is an important tool with which you are addressing the hiring manager personally. It can be somewhat of a daunting task to write a covering letter that really emphasizes the qualities and skills you have to offer to the employer. The covering letter should be well written and succinct and to the point. The importance of a well-written cover letter cannot be emphasized. See below for some ideas which you can use for writing a covering letter.

- Always use professional language and format for a covering letter.

- Mention precisely why you consider yourself suitable for the respective position.

- Try to say something good about the company or firm that you are applying to.

- Never use short forms, always use full words, like do not write etc. but write etcetera, or do not write don't, say do not. This shows that you are different from others and your letter looks professional.

- Use proper grammar in your letter and avoid spelling mistakes and slang at all costs.

- Be to the point but never use bullet points.

- Be brief and succinct and present yourself skillfully.

- Let your letter be unique and stand out from the rest.

- Use some of the skills mentioned in the job posting, this way you can always catch the eye of your recruiter.

- Proofread and edit your letter before sending it. This is your best chance of landing a good job.

- You can also check out some samples on the internet for covering letters. Be careful as there are an awful lot of them, and you should choose the more professional looking ones.

- Never overstate your capabilities and skills. Just keep them to an optimum.

Conclusion

I have tried to highlight the various requirements which the recruiters in the industry look for when recruiting a structural engineer. I have set out requirements at various levels so that a structural engineer working at any level can benefit from these. The requirements set out are practical and easy to achieve, although they may need some investment; but such an investment is worth the money you would spend on these.

Having the knowledge of what to expect will get you ready to face it and build in you the grit required to sustain in the current challenging environment. After reading this book you are now more aware of what is required and how you can get ready to achieve it. With this book, I have tried to precisely answer a question which every structural engineer faces in their lives and that is - 'What Next?' I hope I have answered it and that this book has met your expectations.

Made in the USA
Columbia, SC
16 November 2023